Y0-CXN-841

DATE DUE			

591.59
CAR

Carter, Kyle.

1824

Animals that talk

Quiz # 13053

RL
4.7

WALNUT CANYON ELEM SCHOOL
MOORPARK, CA.

427410 01345   45968B   003

ANIMALS THAT TALK

THINGS ANIMALS DO

Kyle Carter

The Rourke Book Co., Inc.
Vero Beach, Florida 32964

© 1995 The Rourke Book Co., Inc.

All rights reserved. No part of this book
may be reproduced or utilized in any
form or by any means, electronic or
mechanical including photocopying,
recording or by any information storage
and retrieval system without permission
in writing from the publisher.

Edited by Sandra A. Robinson and Pamela J.P. Schroeder

PHOTO CREDITS
All photos © Kyle Carter except page 15 © Tom and Pat Leeson

Library of Congress Cataloging-in-Publication Data

Carter, Kyle, 1949-
    Animals that talk / by Kyle Carter.
    p. cm. — (Things animals do)
  . Includes index.
    ISBN 1-55916-115-9
    1. Animal communications—Juvenile literature. [1. Animal
communication.] I. Title. II. Series: Carter, Kyle, 1949-
Things animals do.
QL776.C37 1995
591.59—dc20   94-47362
                                                       CIP
                                                        AC

Printed in the USA

TABLE OF CONTENTS

ANIMALS THAT "TALK"

People send messages to each other using written and spoken language. Animals send messages to each other, too, but they "talk" in different ways.

An animal may send messages with sounds, scents or **body language.**

People don't always understand the messages one animal is trying to send to another animal. Animal "talk" is full of mysteries.

A bighorn ram's body language—lowering its head—says that it's ready to fight another ram

WHY ANIMALS SEND MESSAGES

Animals of the same **species,** or kind, need to keep in touch. For example, a female bear with cubs may want a nearby male bear to know she is angry and frightened. If she can tell him about her mood—maybe with a growl—she may be able to avoid a dangerous fight with him.

Animals tell each other more than their moods. They tell where they are, where their **territory** is, and when they're looking for a mate.

By lipcurling, a bighorn ram tests the scent—and mood—of a female bighorn

BODY LANGUAGE

When animals use body language, they "talk" with their bodies. Their movements and expressions send a message. An angry bear may growl, but it also shows its teeth and lays back its ears. To another bear, the body language is clear—"Go away!"

Pronghorn antelope and white-tailed deer flash their tails to warn others in the herd of danger. Beavers slap their tails on water to signal danger. A young wolf lies on its back or crawls to say, "You're the boss" to a larger, older wolf.

A young wolf shows respect
by lying on its back

TALKING WITH SCENTS

Many animals, large and small, react to scents. Animal scents are often too faint for human noses. However, some animals have a very good sense of smell. This ability helps them understand messages from other animals.

Animal scents come from body **glands,** droppings and **urine.** For example, tigers and bears mark their territory by leaving scent signposts. Other tigers and bears can tell from the scent who is in the neighborhood.

Members of the dog family, like these young foxes, live in a world where messages are often made with scent

Buzzing rattles help a rattlesnake avoid being trampled by a buffalo or other large animal

*Porpoises stay in touch with undersea chatter that they can hear
even when they're out of sight*

TALKING WITH VOICES

Thousands of kinds of animals "talk" with their voices. Whines, growls, howls, barks, yelps, whistles, chirps, bellows and other calls help animals express their moods and needs.

While many animals **vocalize,** or use their voices, male songbirds are among the few who seem to sing. Their musical notes attract mates and drive away other males.

The song sparrow sings to find a mate

INSECT TALK

Insects don't have voice boxes, but some of them make sounds with their legs or wings. Katydids, for example, vibrate their wings to make a sound that attracts mates.

A honeybee returning from a flower feast can tell other bees where the flowers are by the way it "dances" in the hive.

Cecropia moths and their relatives don't "talk" through sound or dance. A female cecropia says she needs a mate by releasing tiny scent particles into the air.

With just a few days to live, each cecropia moth female uses a powerful scent to attract far-away males

REPTILE AND AMPHIBIAN TALK

Male frogs and toads are terrific callers. Their loud music after dark attracts mates from far away. Each species has a different call, so only females of the caller's species show up.

Reptiles talk in many different ways. Bull, or male, alligators bellow to warn other bulls away and win mates. When the rattlesnake vibrates its rattles, its message is, "Watch out!" The warning helps the snake avoid being trampled by large animals, like us.

Music on a spring night—a male toad's shrill call attracts female toads

BIRD TALK

Birds are famous for their voices. Some of the biggest birds, such as swans and cranes, have some of the loudest calls.

Male songbirds win the hearts of females by singing. Birds also use their voices to send messages of **distress** and give commands to their young, like "Come here!" and "Hide!"

Cranes send messages to their mates by dancing—leaping about and fluttering their wings.

With loud honks, snow geese lead their goslings away from the nest and to the safety of a tundra river

MAMMAL TALK

Next to us, some of our mammal relatives seem to have the most complicated ways of sending messages. The howls of wolves, the mysterious songs of humpback whales and the clicking sounds of porpoises are well known. However, people who study animals don't fully understand what all those interesting sounds mean.

At least one mammal even seems to learn a "foreign language." At the Gorilla Foundation of California, a gorilla named Koko learned to use more than 500 signs in American Sign Language and talked through signs with its trainer.

Glossary

body language (BAH dee LANG gwidg) — motions that carry a message without sounds

distress (dis TRESS) — trouble or danger

glands (GLANDZ) — certain body organs that make liquids

species (SPEE sheez) — within a group of closely-related animals, one certain kind, such as a *brown* bear

territory (TARE ruh tor ee) — the home area that an animal treats as its own

urine (YUR in) — the body's liquid waste

vocalize (VO kul ize) — to make sounds

INDEX